ROBOTICS BASICS

by

Karen Liptak

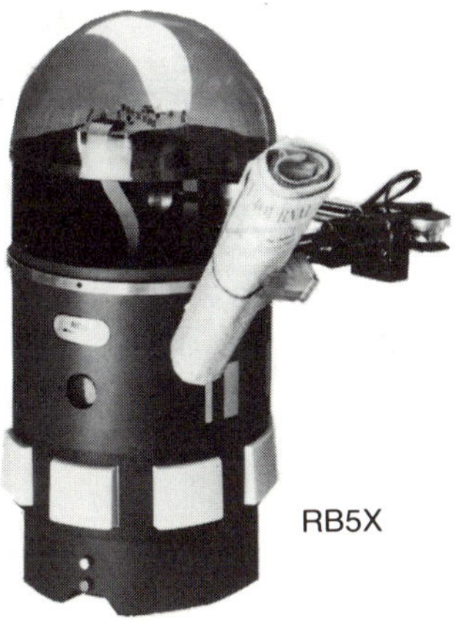

RB5X

Drawings by
Mike Petronella

With Photographs

Created and Produced by
Arvid Knudsen

Prentice-Hall, Inc.

Englewood Cliffs, New Jersey

Photo courtesy of RB Robot

Dedicated to the children of tomorrow
. . . especially Jana.

"Mankind is no longer alone."
Isaac Asimov

Other **High-Tech Basics** Books in Series
COMPUTER BASICS *by Hal Hellman*
VIDEO BASICS *by J. T. Yurko*
HOME COMPUTER BASICS *by Jeffrey Rothfeder*
PHOTOGRAPHY BASICS *by Vick Owens-Knudsen*
COMPUTER PROGRAMMING BASICS *by Lawrence Stevens*
ELECTRONICS BASICS *by Carl Laron*
COMPUTER GRAPHICS BASICS *by Lawrence Stevens*

Printed in the United States of America · J

Prentice-Hall International, Inc., London
Prentice-Hall of Australia, Pty. Ltd., Sydney
Prentice-Hall of Canada, Inc., Toronto
Prentice-Hall of India Private Ltd., New Delhi
Prentice-Hall of Japan, Inc., Tokyo
Prentice-Hall of Southeast Asia Pte. Ltd., Singapore
Whitehall Books Limited, Wellington, New Zealand
Editora Prentice-Hall do Brasil Ltda., Rio de Janeiro

10 9 8 7 6 5 4 3 2 1

Library of Congress Cataloging in Publication Data

Liptak, Karen.
Robotics basics.

Summary: Explains what robots are, and discusses the future role of robots in industry and the home and the relationship between robots and humans.
1. Robotics—Juvenile literature. 2. Robots—Juvenile literature. [1. Robotics. 2. Robots] I. Seiden, Art, ill. II. Title.
TJ211.2.L56 1984 629.8'92 84-4921
ISBN 0-13-782087-9

CONTENTS

INTRODUCTION

Welcome to the Robotic Age

The robots have arrived! They're no longer just science fiction characters. Today they're in our factories, and they're coming into our homes for real. *And what about tomorrow?*

Drawing boards all over the globe are filled with exciting plans for advanced technology robots and "robot families" capable of someday making life richer and easier for everyone alive: a Golden Age on Earth and beyond.

Yet many people eye robots with suspicion. They fear that these marvelous machines will cause widespread starvation as robots take away our jobs and leave people penniless. And they worry that programmable robots with their swift computer brains will someday become so smart that we *will wind up working for* them*!*

Human dreams and nightmares about robots are limited only by the human imagination. But there is one thing we can all agree upon:

The stage has been set.
The trumpets have sounded.
The Robotic Age has just begun.

Karen Liptak

1/ The Origin of Robots

The word "robot" didn't even exist until 1921 when a Czechoslovakian playwright, Karel Capek, took it from the Czech word *robotta* (meaning "forced labor") for use in his play *Rossum's Universal Robots*, or "*R.U.R.*"

The robots in *R.U.R.* are manufactured by humans, whom the robots later turn against.

At the end of *R.U.R.* only one human being still exists on Earth. Robots have taken over the entire planet!

No wonder robots scare some people so much.

Movie Star Robots

Most of us first encounter robots in films, where they've played both bad guys and good guys.

The most famous "good guy" Hollywood robots so far have been C-3PO and R2-D2 from the thrilling *Star Wars* epics. These two metal comedians steal the show as companions to human Luke Skywalker in his battles against our galaxy's greatest villains.

C-3PO, 6 feet tall and a cosmic translator by profession, can speak in 10,000 galactic languages and converse with robots and computers as well as humans. "He" is the gold-metal-skinned humanoid of the pair. (A humanoid is a robot with human characteristics.) And C-3PO, with his polite speech and finicky movements, makes you think of an English butler.

On the other hand, R2-D2 reminds many people of a 4-foot-tall vacuum cleaner on wheels!

R2-D2 can speak in a language of beeps, whistles, and clicks that only other robots and computers can understand. Its main job is servicing starships.

Both robots come with "restraining belts" that keep them faithful to their human owners. (In *Star Wars*, robots call humans Organics. They call themselves Droids.)

Asimov's Rules

The fictional restraining belts worn by R2-D2 and C-3PO in *Star Wars* may have been inspired by the three famous "Laws of Robotics." These "golden rules," written by Isaac Asimov, a well-known science fiction and science writer, are only imaginary. However, they became the basis for many works of science fiction by Mr. Asimov* and others.

Here they are:

1. A robot may not injure a human being, or through inaction allow a human being to come to harm.

2. A robot must obey the orders given it by human beings except where such orders would conflict with the First Law.

3. A robot must protect its own existence as long as such protection does not conflict with the First or Second Law.

As you continue reading *Robotics Basics*, think about what Laws of Robotics you might like to write someday.

*Isaac Asimov's most popular book about robots is *I, Robot*, written in 1950. It contains nine short stories concerning fictional conflicts between humans and robots.

Robots: Their Roots

Most people consider robots the descendants of automatons —machines that do a set number of steps on their own again and again. As far back as 2000 years ago, water-powered automaton figures struck bells at scheduled intervals to help Egyptians keep track of time.

Centuries later, aristocrats collected small automatons to amuse their guests. These devices, operated by springs and/or gears, might play the piano, sharpen knives, light cigars, pour dinner drinks, or even write notes.

The Mechanical Writer, created by an eighteenth-century father-son team, Pierre and Henri-Louis Jacquet-Droz, writes messages of up to forty letters each. To change the message, you simply adjust the gears hidden inside The Mechanical Writer's chest.

Today, many amazing toy automatons are available for people to buy. Most are powered by winding a key of some sort, which in turn tightens a spring inside. Then as the spring uncoils, the figure moves.

In factories, automatic machinery powered mainly by electricity speeds production and saves labor costs for employers.

Electrical Robots

Electricity first became widely available as a source of power in the early twentieth century. In 1939, two electrical robots—Elektro and Sparko—were smash hits at the New York World's Fair.

Elektro, the humanoid of the pair, stands 7 feet tall and weighs 260 pounds. His nervous system of motors, levers, gears, and chains allows "him" to walk, salute, distinguish colors, say 77 words, and do other crowd-pleasing feats. Most impressive of all, Elektro seems to obey spoken commands from humans.

Actually, Elektro responds to the timing and number of syllables he hears, rather than to words. In turn, Elektro gives commands to Sparko, a robot dog that barks and wags its tail.

Sparko has photoelectric cells that attract it to light: It can follow the path of a flashlight moved in front of it. Sparko has another special feature—a built-in heat sensor—that makes it hard

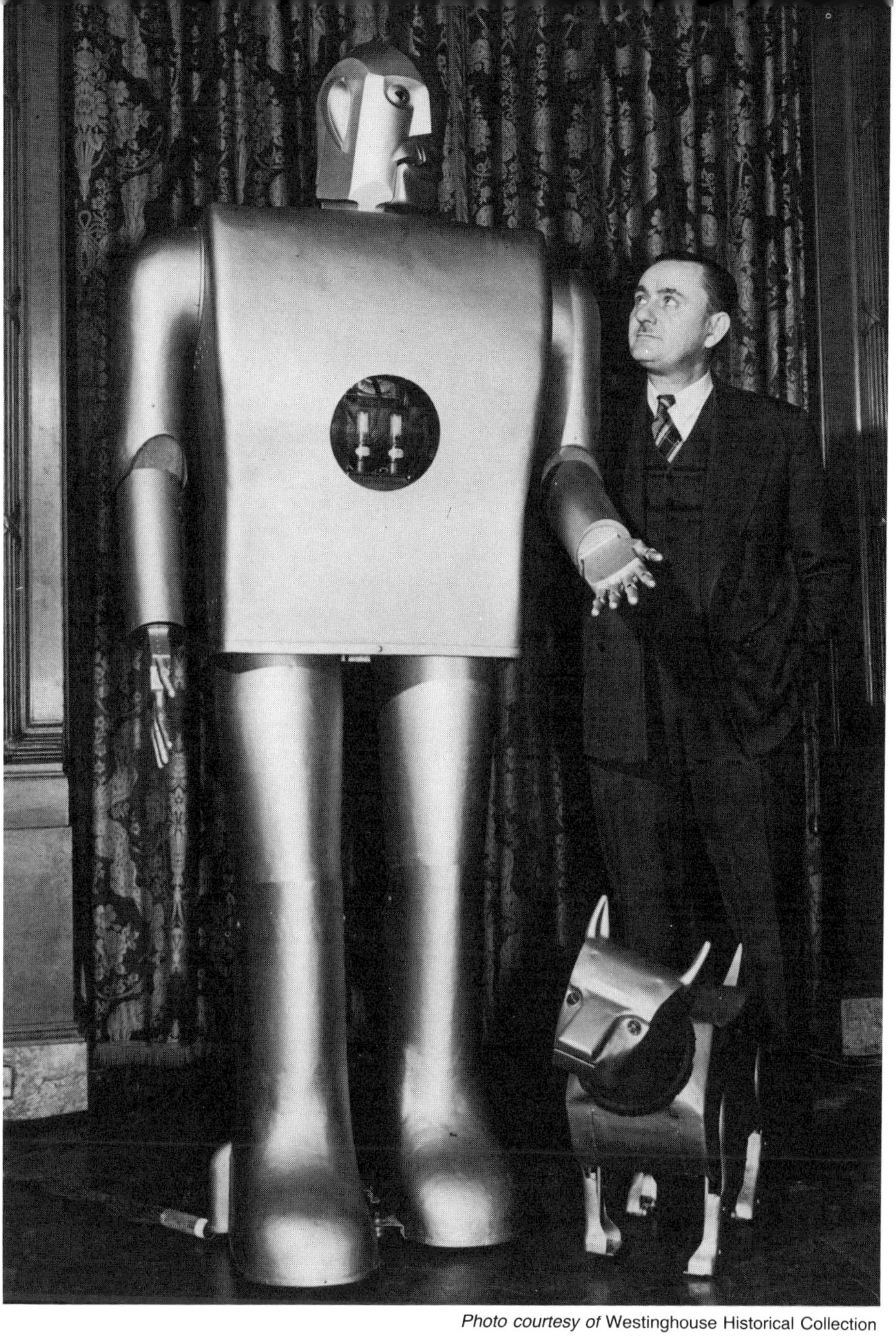

Photo courtesy of Westinghouse Historical Collection

Elektro and Sparko with J.M. Barnett, creator and tutor of the mechanical duo.

to ignore. When visitors approach Sparko, the heat from their legs activates the electrical animal's heat sensors and Sparko trails after them.

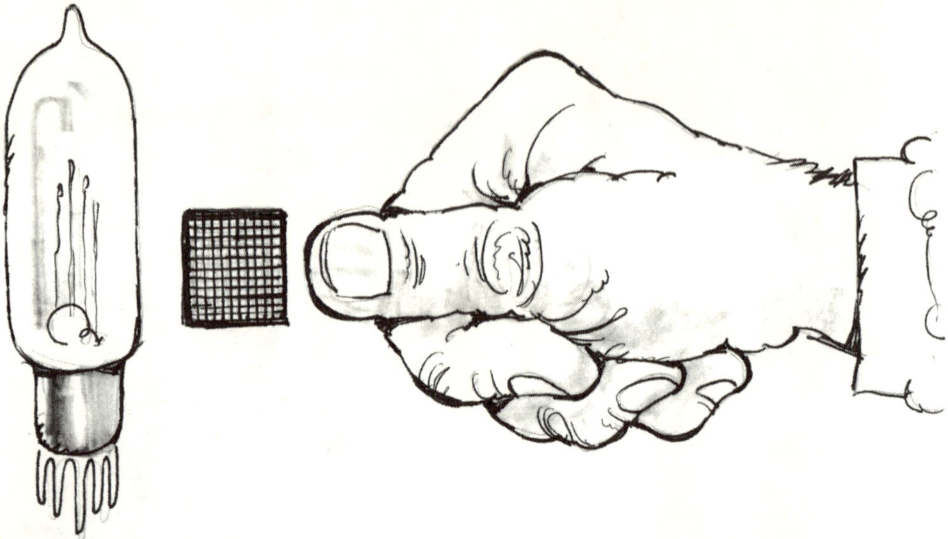

2/ Programmable Robots

Today's robots, while still primitive, are far smarter than ever before. They can perform not just one set of steps but many "programs," depending on the instructions stored in their electronic computer brains.

A robot's computer brain tells its motors what to do. (The word *computer* comes from Latin and means to count or figure out answers to problems.)

Plans for the first computer were drawn up in the nineteenth century, but it wasn't until the 1930s that the technology existed for general purpose computers to be built. At first they were operated by electrical switches. Then the growth of the electronics industry led the way to the world's first all-electronic computer in 1946. ENIAC was powered by 18,000 electronic vacuum tubes through which electrons flowed to form an electric current. It filled a space big enough for a two-car garage and could do 5,000 mathematical calculations per second.

Does that seem like a lot to you?

In its heyday, ENIAC was 1000 times faster than anything else around. But compared to current computers, ENIAC is a dunce.

Photo courtesy of Smithsonian Institution

(Today's most powerful computer, CRAY-I, can process 200 million calculations per second!) And ENIAC had a very annoying habit. It averaged one breakdown every seven minutes.

A later effort by ENIAC's inventors broke down only two or three times a day. Called UNIVAC, this 8-foot-high vacuum-tube computer went to work for the U.S. Census Bureau in 1951. Yet even though it solved some major shortcomings in ENIAC, UNIVAC was soon to be outdated by a "giant" of teeny proportions.

Birth of Transistors

Imagine a world without transistors. No pocket radios, calculators, or tape recorders. No portable televisions. No family video games. And certainly no personal home computers small enough and cheap enough for the average person to own.

Three scientists at Bell Labs invented the transistor shortly after ENIAC went into operation. And like a vacuum tube, a transistor controls the flow of electric current in electronic devices.

However, transistors are undeniably superior to vacuum tubes. They are smaller, faster, safer, and more dependable than their bulky ancestors, and they can be produced for a lot less money!

Within a few years, Leprechaun, the world's first completely transistorized or solid state computer (because current flows through solid material, not a vacuum) was built. Each component in Leprechaun had to be connected to other components by tangles of wires.

Then came another major breakthrough. Scientists discovered how to etch the wiring directly onto a piece of silicon. Now computers could be smaller and less cumbersome than ever!

It's All in the Chips

By the late 1950s, computer engineers had learned how to etch the wiring plus the transistors themselves on tiny pieces of silicon "chips." Today, hundreds of thousands of transistors and wires can be etched on a single chip about the size of a pinhead!

Once microscopic silicon chips became a reality, robots of manageable size could be built with programmable computer brains. These brains enable the robots to do a variety of tasks, rather than just one thing again and again.

The gap between automatons that could do only one job and robots was widening. A new breed of machine had been created!

What Is a Robot Today?

The definition of a robot is "growing up" along with the creation itself. The more sophisticated we make this machine, the more we need to change our meaning for robot to fit what it becomes.

For today, one robot manufacturer says "a robot is a mechanical device that can be taught to do a variety of complex jobs." Another robot maker describes a robot as "really just a computer, but one that is able to sense and react to its environment."

And for tomorrow? What will our definition of robot be then?

Raj Reddy, the visionary director of the Artificial Intelligence Laboratory at Carnegie-Mellon University, calls a robot a "system that can see, hear, walk, think, and act with its environment."

Robots that *really* think? By that definition, no robot exists on Earth *yet*!

Robots for Hire/3

Robots have crossed over from fiction into fact most obviously on assembly lines in factories. From 60,000 to 80,000 industrial robots are in use today. And more are being "hired" daily.

Industrial robots, controlled by computer brains that let them do many tasks, depending on the programs stored in their electronic memories, are used mostly by the auto industry. (Spot welding and painting car bodies are their two main jobs.)

Since car manufacturers come out with new-model autos every year, single-task automatons had to be scrapped every twelve months. But programmable robots are far more flexible—they'll work on a different model car every week if you program them to do so.

Robots are also currently being used in the manufacture of bricks, glass, metal, plastics, jet planes, and more.

New, advanced robots are starting to do precision work on electronics and small appliance assembly lines, too. Transistor radios, razors, kitchen counter mixers—robots are beginning to make them all.

Photo courtesy of Cincinnati Milacron Co.

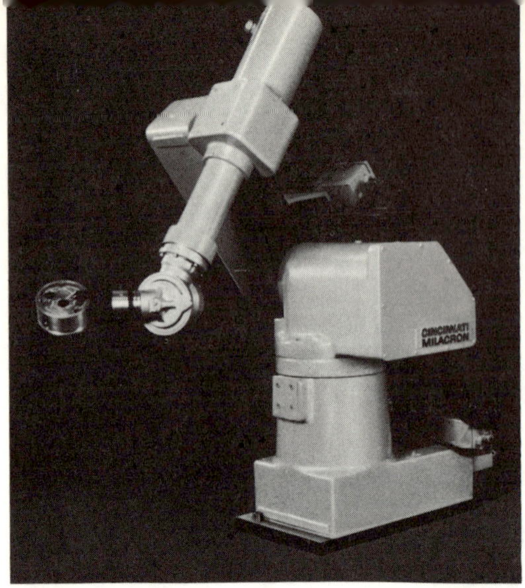

Cincinnati T3 industrial robot models. Note the gripper for handling materials.

This is a Robot?

Don't expect industrial robots to look human. Don't even expect them to resemble R2-D2. In fact, most could pass for huge dentists' drills!

The distinguishing feature on most industrial robots is a mechanical arm with hands or grippers that lets the robot interact with objects on the assembly line. (Hooks, spatulas, scoops, electromagnets, and vacuum cups are some other "end effectors" robot arms may have.)

While industrial robots are *controlled* by the instructions programmed into their computers, they are *powered* most often by electricity. Pneumatic (air) pressure is also used. And for heavy chores, robots receive their power from hydraulic, or fluid, pressure.

Why Employers Choose Robots

Robots are very popular with their bosses. They are efficient, tireless workers who don't stop for lunch, coffee breaks, or even to use the bathroom. They'll work three shifts, round the clock, and still be ready for a full day's labor the next morning.

Robots won't ever strike or blame the poor quality of their work on personal problems. They won't complain—no matter how dull, dangerous, or dirty a job they've been assigned. They will just do as they have been programmed to do until they either break down (which shouldn't be too often) or get reprogrammed for a new task.

And reprogramming an industrial robot may take a matter of minutes or hours, while it could take many weeks, months, or even years to retrain a human worker.

Here's the low-down on robots in dollars and cents: A robot welder may cost its owner $30,000 to install. It should then pay for itself in three to four years and not require major repair or replacement for approximately eight years. The robot may average out to cost its owner $1 an hour for its services. A human worker at the same job could cost an employer $9–$10 an hour or more at current rates.

Photo courtesy of Unimation Corp.

Industrial robots "employed" in glass factory.

Photo courtesy of Unimation Corp.

Industrial robots busy at work in auto factory.

Friends or Foes?

When "Clyde the Claw," a robot in a Chicago auto plant, broke down, its fellow human workers reportedly organized a get-well party and showered its pedestal with flowers and cards.

But not all workers feel so chummy with robots. Many worry about recent estimates that robots could replace up to 75 percent of all factory jobs in the next decade. "How can we feel good when our livelihoods are being threatened?" they ask.

Others are more optimistic. A recent congressional report stated that "robots will create new jobs and raise productivity and with that the material rewards to employers and employees alike." And one robot researcher points out that robots will only be kept performing at top efficiency by a team of people all working together. This team will include electrical engineers, programmers, and operators, as well as repair and maintenance specialists.

Japan's Example

Once called "The Land of Tradition," Japan is today better known as "The Island of Robots."

While robot technology in the United States is equal to that in Japan, the Japanese lead the world in robot applications. More than 35,000 metal workers are currently helping Japan's economy to boom, compared to 7000 industrial robots in the United States.

Japan has therefore been forced to deal with the problem of robots replacing humans much sooner than we. Japanese trade union agreements now guarantee that workers cannot be discharged or demoted when robots enter their workplace. Rather, they must be retrained in salable electronics skills.

In the United States, concerned companies like IBM also have policies for reskilling workers replaced by robots.

But problems still remain.

Reports from Japan indicate that middle-aged workers sometimes feel sidelined when their previous skills become obsolete and they are given new jobs to learn.

"My hands could cry," said one displaced Japanese factory worker in a recent TV documentary.

The answer to the problem may rest in the process of robotization being done very gradually, perhaps over generations rather than a few short years.

Photo courtesy of IBM

Multiple flash exposure photo of IBM 7365 robotic arm used in electronics, automotive, aerospace and appliance industries.

The Next Wave

Most assembly-line robots are hefty machines weighing in at 1500 pounds or so. But Unimation's advanced programmable PUMA robot tips the scales at less than 100 pounds. No larger than a human being, PUMA was made for General Motors, which specified that the robot be able to handle a weight capacity of 5 pounds.

Why only 5 pounds? Because General Motors' research showed that 90 percent of all the parts in an auto weigh no more than that.

Another advanced industrial robot—Cincinnati Milacron's T3—can monitor its own problems and in some cases fix its own parts when they break down.

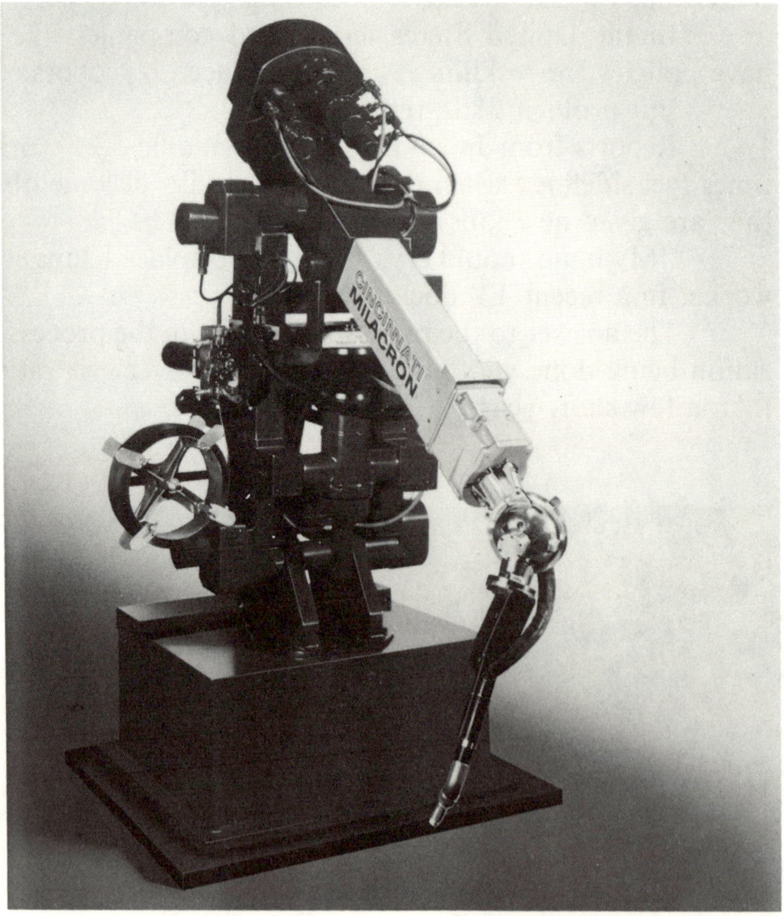

Photo courtesy of Cincinnati Milacron Co.

Cincinnati Milacron T3 industrial robot with arc welding equipment.

Personal Home Robots: The Next Great Home Appliance? / 4

"Can anyone really envision the year 2000 without robots around the house?" asks the founder of a personal home robot company.

Suddenly, people everywhere are infatuated with these exciting computers on wheels that can amuse you, educate you, and even do a few useful chores around the house.

They are not to be confused with yesteryear's delightful automatons. These robots are meant to be much more than toys.

Fun, Games, and More

Three of today's best-known personal home robots are B.O.B. (Brains On Board), RB5X, and HERO I. They range in price from $1500 to $2500, and each has a built-in computer.

Optional "Androwagon" lets B.O.B. transport school supplies around house.

Photo courtesy of Androbot

B.O.B. is a 3-foot-high slant-wheeled life of the party who can astonish guests by telling them party jokes and fetching them drinks dispensed from a special refrigerator accessory.

But B.O.B.'s manufacturers have more serious plans in store for their futuristic-looking family robot. Software is being planned to allow B.O.B. to teach languages, math, geometry, and other subjects, as well as wake up sleepers, take phone messages, and watch over the total security of your home.

B.O.B. can already be programmed to sound an alarm when its sensors detect an intruder. But someday it will also be able to alert the police by some kind of direct communications interlink.

Also being planned is a way for B.O.B. to respond to spoken commands. Another home robot scheduled to have a primitive voice-recognition system is RB5X.

Robot Fire-Fighter

RB5X will soon be able to fight home fires, too, thanks to a Colorado high-school student named Marc Turner. Marc won first place in the Engineering Division of the International High School Fair in Albuquerque, New Mexico, competing against students from twelve nations.

Marc's winning entry—a system for detecting and extinguishing fires—will become one of the RB5X's security features. When a fire is detected, RB5X will sound an alarm, as well as squirt water on the flames until they're out.

Another interesting option for RB5X is a vacuum attachment. The robot will be able to map out an area to be cleaned and then get to work. Its sonar equipment will keep it from bumping into furniture. But RB5X still has a long way to go before its intelligence allows it to know when a certain spot needs a second going over.

Robot from a Kit

HERO I is a personal robot you can buy already assembled or in a kit form with 1200 parts you put together by yourself!

And what do you get for your efforts? An adventurous little friend with a computer brain that zips along at a million operations per second and a mechanical arm capable of moving in seven different ways. We call these motions a robot's "degrees of freedom."

Most industrial robots used in simple factory jobs have from three to six degrees of freedom. The human arm has twenty-seven. (RB5X's fold-up arm moves in five directions and can be programmed to reach around and shut RB5X off!)

Powered by four rechargeable lantern batteries, HERO I has sensing devices that keep it from crashing into tables and lamps, and a speech synthesizer that enables it to reproduce any sound it hears.

HERO I can be programmed up to a year in advance, which means you can instruct it to sing "Happy Birthday" to you twelve months from now, even if nobody else in the whole universe remembers!

Limitations

Of course, today's home robots still have far to go before they're real domestic servants. Our simplest chores remain devilishly complex to them.

You still can't program a robot to take out the garbage and be sure that it won't leave the trash behind and empty the family cat into the refuse truck instead!

And modern homes—with the people and objects in every room constantly rearranging—are much too complex for a robot of today's generation to get around in safety. Industrial robots have an easier time of it, since they're usually bolted down and programmed to deal with only a limited universe of movements and space.

The "universe" an average home presents to a mobile robot is enormous to say the least!

Then, too, robots won't be truly efficient domestics until they can manipulate regular-sized objects (HERO I holds up to five ounces and RB5X, up to twelve) and are able to tell upon contact how much pressure to apply to an item or surface. Before that time, washing dishes and cleaning windows are two chores you had better not ask a home robot to do.

The technology we would need to turn current home robots into smart servants either isn't available yet or is too expensive to be practical for today.

But there's always tomorrow. . . .

Users Needed/Potential Unlimited

You can compare personal home robots today with personal home computers when they first came out in the late 1970s. People wondered back then what more this new electronic "toy" could do besides balance the checkbook.

Then computer-users began creating the software, or programs, that made computers increasingly versatile and indispensable as well! (*Time* magazine recently reported that by 1986 Carnegie-Mellon University expects to require all students to have their own personal computers.) Some other colleges have already made this requirement.

"We are counting on our users to come up with the software," the manufacturer of one personal home robot confesses. "It will be like having hundreds of programmers on our payroll."

Will you be one of them?

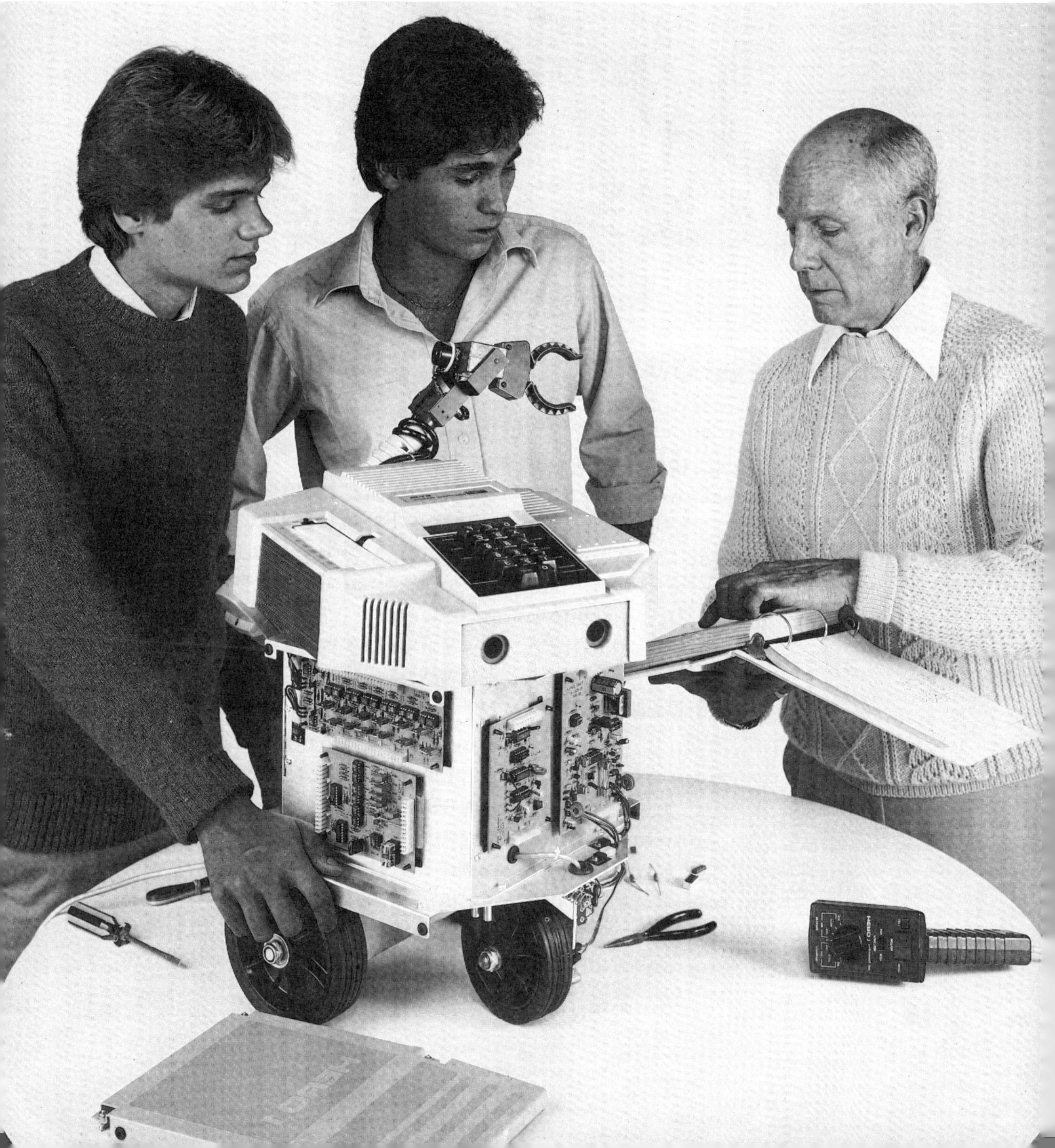

5/ Robots Busy Everywhere

In Australia, robots are helping farmers shear sheep.

In Israel, they're picking avocados.

In a Hollywood movie studio, a computerized robot camera operator takes six seconds to photograph a special effect that once took four people three hours to complete. And in both the United States and other countries, slightly larger than human-size robots are directing traffic whizzing by at dangerous intersections.

Every day, people are putting robots to work in new ways.

Bomb Squad Robots

Six years ago, the British Army in Northern Ireland enlisted its first remote-controlled robot vehicle to remove and dispose of terrorist bombs. These "wheelbarrows," as they are now called, number 600 today and do a variety of dangerous jobs for police departments and rescue teams around the globe.

Most wheelbarrows come equipped with a mechanical arm that accepts an assortment of accessories, depending on the mission. Some can climb stairs and sidewalk curbs. And a wheelbarrow armed with a 360-degree panning camera and a microphone and speaker can save many lives by acting as a bulletproof communications link between negotiators and gun-toting terrorists.

Mailmobiles

"Little Go Beep," "The Flasher," "Happy Honker," "Beeping Tom," and "Archie Bumper" are all names employees in different companies have affectionately given their Mailmobiles: 4-foot-high battery-operated office robots.

These mechanical mail carriers move along invisible phosphorus paths sprayed on the floor and make a programmed number of twenty-second stops. When their time is up, Mailmobiles beep a warning and then continue on their routes—even taking the elevator should the occasion arise!

The most advanced Mailmobiles have sensing devices enabling them to stop when an obstacle is in their path. Others halt upon contact.

Mailmobile "Little Go Beep" shown decorated for company picnic.

Photo courtesy of Peerless Insurance Company

"Hot Cell" Robots

Nuclear energy is valuable for producing electricity and is used in medicine and science. But the radiation from nuclear materials can damage human tissues. Therefore, nuclear scientists count on robots to help them do their research and their work.

Mobot was one of the first robots created for this hazardous mission. It has long, flexible hands that can pour radioactive liquids from one container to another under the careful directions of a human operator who sits at a control panel and watches a monitor showing photos of what Mobot's camera eyes "see."

The control panel could be behind a sealed glass window several feet away from Mobot or it could be on Earth while the robot worked elsewhere in the solar system.

Another hot cell helper, the robotic "slave arm," is electronically connected to a special "master arm" device worn on the arm of a person sitting at the control panel somewhere. When the person moves her arm, the slave arm mimics each action!

Both Mobot and slave arms require constant human supervision.. They are not programmable, but rather operate by remote control. For today they exist in a twilight zone—not quite robots but more than automatons.

Sick Robots

SIM ONE often needs a doctor. This robot android—a robot in human form—who does his best work lying down is an imitation

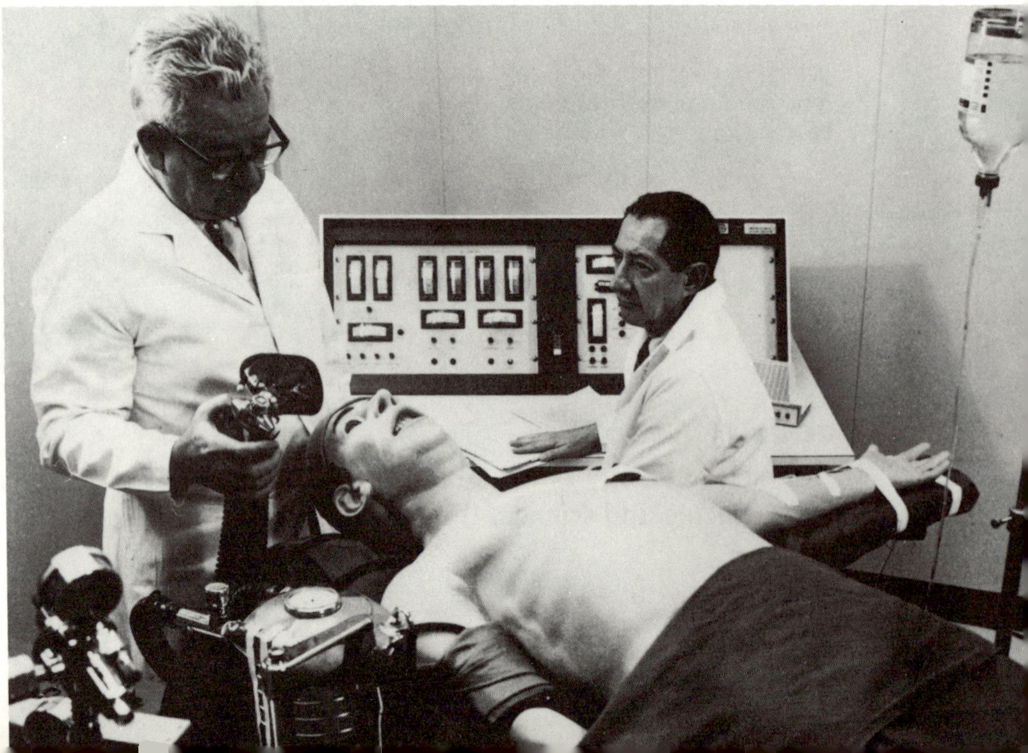

patient at the University of Southern California. He's got a heartbeat, pulse, plastic skin that seems real, and a computer hidden within his bed that is able to be programmed for SIM ONE to suffer from a variety of heart diseases that the medical students at USC try to cure!

Future models of SIM ONE are being planned with wrinkles on his brow that will grow deeper when he receives certain kinds of medicine!

Robots Take a Dive

A master diver can work well only at 1000 feet. However, a robot diver can descend 2000 feet underwater with ease. And a robot diver can work 24 hours a day, while a human diver is efficient for only 10 minutes at a time. Just as attractive is a robot's price. It might cost $7,000 a day, while an expert diver could cost an employer $25,000 daily.

Remote-controlled vehicles and slave arms are already being used to search for offshore oil and gas, and in the construction and repair of offshore drilling rigs. Salvaging sunken ships is another current underwater robot application.

Robots to the Stars

Robots would make better astronauts than Earthlings in many ways. They wouldn't require bulky life-support systems, and they could survive for long periods without food, water, sleep, or human society. Robots can also more effectively withstand solar radiation than our more fragile human stock.

Space robots—controlled remotely from Earth but capable of some on-board navigational chores, taking photos, and recharging themselves—have already flown to places humans have yet to go.

In 1976, two spidery United States robot landers, Viking I and II, detached from their Orbiters to touch down on Mars and immediately began sending to Earth the results of numerous experiments performed in their portable scientific laboratories.

After centuries of fantasizing about life on the "Red Planet," the moment of truth was closer than ever!

◄ SIM ONE at work
Photo courtesy of Univ. of Southern California

Photo courtesy of NASA/JET Propulsion Laboratories

Model of Viking Lander in a Mars simulation laboratory. The two cylinders on top are survey cameras.

Each Viking Lander's mechanical arm skillfully scooped up Martian soil and tested it for many qualities, including any signs of life as we know it on Earth. The results were negative.

Though the machines were given a life expectancy of only a few months, Viking I continued to thrill scientists worldwide with weekly radio reports from Mars until late 1982.

Russia, too, has space robots. It was their Venera 13 and Venera 14 Landers that sent us a stunning gallery of photos taken from beneath Venus's bright orange sky. But Venus's surface temperature is so intensely hot (850° Fahrenheit) that these brave metal explorers lasted on it for only two hours before the heat destroyed them.

In June 1983 a robot spacecraft, Pioneer 10, became the first Earthmade object to go beyond our solar system's most distant planet (Neptune at the time). Unless it is intercepted, Pioneer 10 won't approach another star until some 10,000 earthyears from today, when it comes approximately 23 trillion miles near Barnard's Star.

Perhaps beings on planets around that far-away sun will be the first to read the "postcard to posterity" on Pioneer 10's side: a plaque depicting a male and female human figure and a map showing Earth's position in space.

Shuttle Robots

Robotic arms are being used in satellite retrieval and repair missions on space shuttle flights. The first test of a shuttle arm was made on a Challenger space flight in 1983.

Sally Ride, the United States' first woman astronaut, helped control the robot arm to release a 15-foot package of scientific instruments from Challenger's cargo hold and deposit it into a separate orbit.

Hours later, the robot arm was commanded to pluck the package back from space in good shape.

The test was a huge success. And in April of 1984, the first service call in space was officially made! Astronaut Terry Hart operated Challenger's 50-foot robot arm to snatch the world's first floating solar observatory, Solar Max, into its cargo bay for repair. Later, the robot arm was directed to return the fixed satellite to orbit.

A new era of space age salvage had begun!

A Robot Draft?

"Why don't the robots fight our wars and we all stay home?" suggests one robotics expert. Actually, robot soldiers (if only we could create them!) make great sense. They wouldn't need food or shelter, and think of the lives they could save!

Another possible war machine—one on the drawing boards right now—is a sphere-shaped robot named Omnibot that can be parachuted into any dangerous situation. Once on the ground, Omnibot will be able to maneuver into the correct position, then open up to reveal arms, a TV camera, and various sensing devices.

Should danger suddenly threaten, Omnibot will be programmed to quickly fold up into a ball again and split fast!

"Showbots"

These are seen in shopping centers. Some are androids—human-looking in form. Others are more strangely shaped. Fubar D. Robot, a mobile blue-and-white conglomeration of chrome, plastic, and mirrors has shimmering gold accordion arms, a head that turns 360 degrees, and a sense of humor.

Ad for Fubar D. Robot, shopping mall "Showbot" and guest star of T.V. series "Silver Spoons."

Here's how Fubar explains its origins: "My mother was a washing machine, and my father was a fourteen-speed blender."

Most Showbots are remote-controlled, but others can be programmed in advance. In any case, they can be considered mechanical performers, rather than real robot workers.

The fascinating computer-operated mechanisms, called Audio-Animatronics, at Disneyland and Disney World also perform rather than work. They, too, are Showbots, not bona fide robots of today.

How to Teach and Improve Robots / 6

As you already have learned, true robots have computer brains that control what a robot does by the programs or software fed into them.

These brains generally have several tiny silicon chip packages all plugged into a plastic circuit board. One or more chips (called microprocessors) are in charge of the computer's ability to process data. Other chips store data and new commands. And others hold a permanent control program that coordinates all the computer's other chips.

A circuit board may also have interface chips plugged into it that give the computer brain the capability of linking up with other devices, such as sensors.

Expandable circuit boards have room for additional chips to be added as new device options become available.

Programming a Robot

RB5X personal home robots have a radio communication option that lets one robot communicate what it has learned to another robot. But most of today's robots are taught by human beings.

A programmer teaches, or programs, a robot by typing a series of instructions on a keyboard or keypad, either located "on-board" the robot or on an external computer.

Each instruction then gets translated into electrical charges that are stored in the robot's memory until needed.

Robot operators can run a program from the computer's keyboard or keypad. Or else, in simpler robots, teaching pendants are used. These pendants most resemble handheld calculators.

There are many robot "languages" in use today. The easier they make programming and operating a robot, the more "user friendly" we say they are.

"Atom" the robot waiter, and its creator, Philip Baron

Photo courtesy of Saul Keslowitz

Robots That Listen and Talk

The most user friendly robots of all would be those that respond to the spoken word from many people. Their existence would spell an end to keyboards and keypads. Your voiced wishes would be a robot's commands.

But to truly hear, robots would have to adjust to many different voices, which is no easy chore. The world is filled with vastly dissimilar accents, grammatical structures, and pronunciations. Even the same person seldom says the same word the same way twice.

For a robot to recognize speech, it must have stored in its memory a wealth of data about spelling, pronunciation, and grammar, as well as the meaning of words. (Some machines already do exist that respond to spoken commands, but only from one or two people, and then only if the speakers exaggerate each pause between words.)

Robot Vision

In its most primitive form, "vision" is allowing some industrial robots to act as inspectors on assembly lines. When a robot's program signals that a certain part is flawed, an airjet blasts that object from the assembly line into a reject pile.

But robots don't "see" like humans do. We know what we're seeing by comparing objects and scenes before our eyes with our past knowledge, ideas, experiences, and associations accumulated over many years. Robots lack this human ability.

"Seeing" for a robot means having its TV camera eyes scan what it is viewing and then converting the picture into an electronic map of numerous black-and-white dots. The dots receive numerical brightness values that are fed into the robot's computer.

In the robot's computer memory are measurements for objects the robot is expected to know. The new brightness values are compared with the stored measurements to identify what the robot is now seeing.

The more memory a computer contains, the more images and shapes the robot can recognize. Yet today's most advanced robots are still slowpokes when it comes to seeing. They take minutes to identify the simplest shapes.

However, some robotics researchers believe that within fifty years robots will see as well as humans. Perhaps even better.

Human biology limits us to viewing the world only at wavelengths within a certain range of the electromagnetic spectrum. But robots with supersensitive vision may someday see beyond our limited vision into areas such as the ultraviolet and infrared wavelengths. Microscopic, as well as telescopic, robot vision may also be possible within the next few decades.

In fact, there may even come a time when robot vision is so superior to ours that only robot drivers will be permitted on our streets!

Watch Out!

Factory robots usually don't have to worry about bumping into obstacles because most of them are securely fastened in place. But any robot that's mobile needs to know when something or someone is in its path.

Sonar sensors, dependent on sound waves that bounce off objects, are most often used today to let a robot know when to stop or swerve. And for tomorrow obstacle-avoidance laser systems are being developed on some prototypes of space rovers destined to explore the surfaces of the moon and Mars.

The Perfect Touch

Unless we're infants, ill, or handicapped, we can usually tell with ease how hard or softly to handle an object. But robots still have far to go before their tactile sensors (also called artificial skin) are anywhere near as sophisticated as our sense of touch.

Some of today's robots have primitive touch devices called strain gauges connected to their hands. A strain gauge has springs that can measure the force needed by the robot's grippers to hold an object. The power to the robot's hand is then adjusted so that it won't crush the object or grasp at empty air.

More sensitive touch sensors are currently being worked on so robots can be trusted to do more of the work now earmarked for human beings.

Robots That Think?

When it comes to thinking and making decisions on their own, today's robots are no Einsteins. Although we've managed to reduce the 1950s bulky transistors to silicon chips no bigger than matchheads, we've yet to build a robot able to think like a 3-year-old child.

But the future is approaching. And Artificial Intelligence (the science of getting machines to do thinking tasks) is attracting some of the world's most brilliant minds. The basic problem they must tackle is how to get a robot's computer brain to imitate a human's, when the human brain is still such a mystery.

Humans possess an uncanny ability to organize all kinds of diverse data they receive through their senses and to synthesize the information they learn.

Robots store the data their computer brains receive in the form of lists. When they are asked questions, they search these lists to find the answers they seek. This can sometimes be a very time-consuming process.

If you're asked a real stumper of a question, chances are that you'll know enough to say "I can't help you" fairly fast.

But ask a computer a question, even one for which it has no answer, and it can spend hours searching through its lists until it finally reports back that it's bewildered, too!

Someday, an extensive "world model" of possible responses to potential problems may help robots act more on their own. Such a world model would require enormous circuit boards, capable of storing huge amounts of data. Or else, biochips.

Tomorrow's Earth-shaker?

Some researchers worry that the tiny circuits used in smarter and smarter computers may someday squeeze together so closely that their silicon chips overheat and melt.

And they theorize that biochips—small packages of computer circuits assembled from organic molecules—may be the wave of the future. Should biochips ever be produced, a whole new family of computers will spring up! A family far faster, smaller, and cheaper than even today's most advanced "state of the art" models.

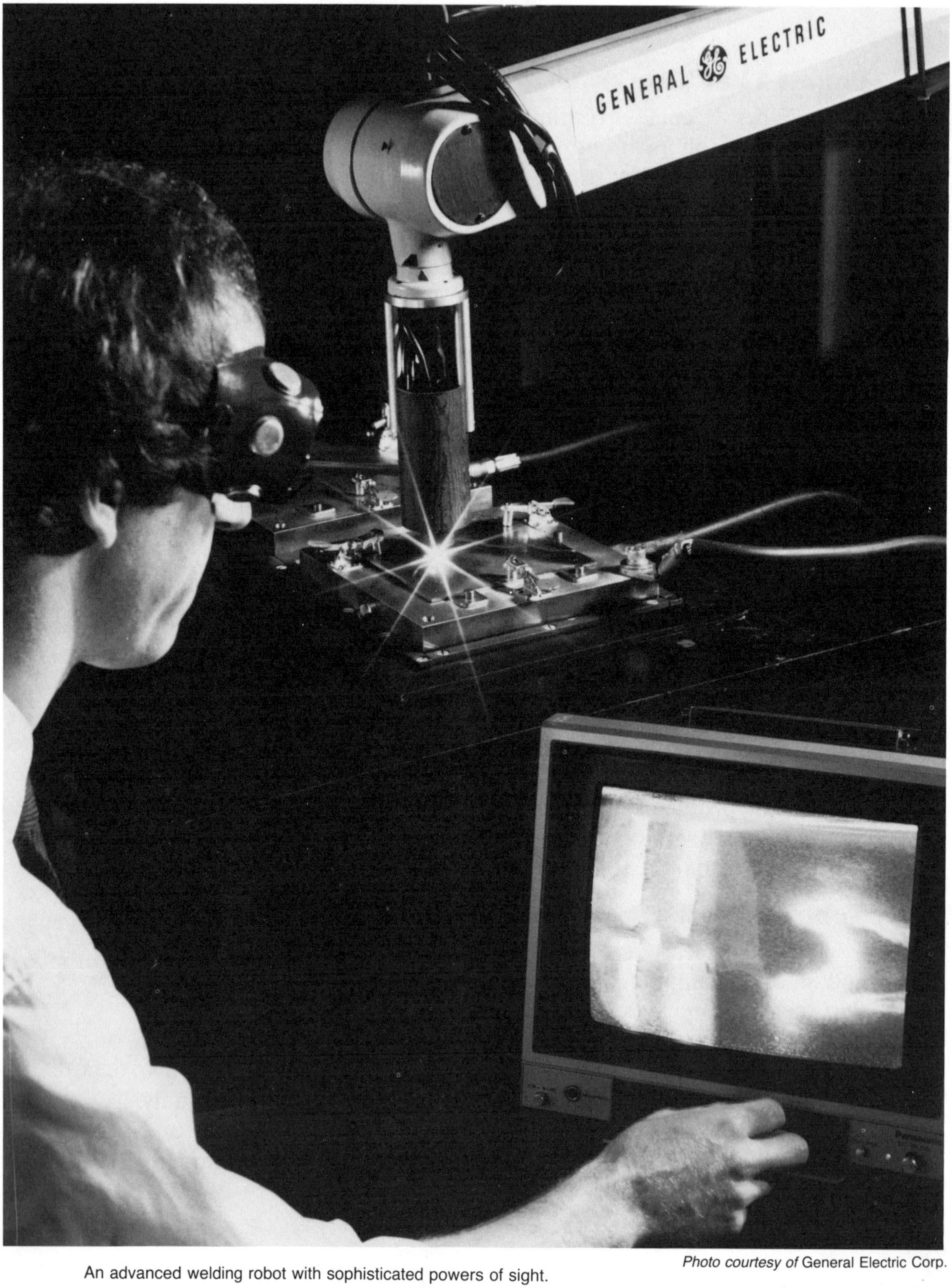

An advanced welding robot with sophisticated powers of sight.

Photo courtesy of General Electric Corp.

New Robot Occupations/7

The Japanese are already creating a McDonald's-like robot that cooks burgers, pours drinks, and makes change. And at a pub in Farmingdale, Long Island, Atom—a 135-pound robot with an on-board computer—greets customers and brings drinks to their tables.

Japanese home builders are starting to use robots to make prefabricated houses. Japanese clothing manufacturers are developing robots that can design, cut, sew, finish, and inspect apparel. Robotic insecticide spraying, egg inspection, and fertilizer spreading are scheduled to begin soon in Japan. By 1990, the Japanese also expect to have in operation robot nurses, guides for the blind, and street sweepers.

Robot helps Quebec cops nab gunman

United Press International

QUEBEC — A man armed with a 20-gauge shotgun wounded two people yesterday before barricading himself in a house ringed by police who used a robot dubbed "Inspector Hercule Poirot" as an intermediary.

The incident occurred a day after a gunman shot his way into the nearby Quebec legislature, killing three people, wounding 13, taking a hostage, and holding hundreds of police at bay for nearly five hours before surrendering.

That suspect, Canadian Forces Cpl. Denis Lortie, was charged yesterday with three counts of first-degree murder.

Quebec Constable Yvan Pinel said the latest shooting may have been motivated by the attack on the legislature. Police identified the suspect as Jean-Claude Nadeau, 39.

They said police had drawn up a personality profile to aid in negotiations, which were conducted through a robot.

Sharpshooters wearing bulletproof vests took up stations around the two-story row house in the historic old town of Quebec. Police cleared near-

News item in recent edition of Northern New Jersey's Bergen Record.

Primitive "security robots" already exist that can detect an intruder and sound an alarm. Someday these mechanical sentries may also carry, or have built in, devices that can paralyze criminals. And more advanced models of today's factory firefighting robots may eventually be able to smother raging forest fires with chemicals, while humans watch the blaze from a safe distance miles away.

Robot Super Servants

From today's personal home robots may emerge a whole menagerie of robot superservants: cooks, babysitters, gardeners, and more, as well as robotic arm attachments to improve our present labor-saving devices. A washing machine with a robotic arm could not only wash clothes, but also put them in and take them out of the machine.

Robot assistants may someday escort you around, reminding you about appointments and taking dictation for letters to be sent via electronic mail. They could also note your daily data requests. Then they might search through their internal data bases or the data bases on the family's home computer.

Or else, via a radio communications hook-up, they might contact any number of electronic data bases around the globe to ferret out whatever information you desire. (Today, in the United States alone, there are 1450 such bases.)

Last, but certainly not least, it is conceivable to someday put a family robot in charge of your entire kitchen operation. The robot could plan the meals, order the food from local roboticized supermarkets, as well as cook up gourmet specialties every night of the week and even wash the dishes when everyone's done.

Eventually, personal robots (whether a family of them or one all-around home appliance model) may be the norm, and be as common in tomorrow's home as a television set is in today's.

"Miracles" for the Handicapped

Exoskeletons are robotic metal frames that people can be strapped into for extra strength. Hardiman, an exoskeleton from General Electric, allows its wearer to lift up to 1500 pounds—much more than an average person can.

Exoskeletons, like their near-relatives' slave arms, are already being used in devices that help handicapped people better control their bodies. Artificial hands now exist that are controlled by muscles in the wearer's arm. Eventually, scientists hope to develop devices that receive orders directly from the brain.

And three aerospace engineers are presently working on a voice-activated robotic system for Jamie, a 3-year-old boy born without arms and legs. Their device, which will combine a high technology computer brain with small powerful motors (some adapted from NASA* projects), will someday be capable of picking Jamie up, adjusting his height, sitting him at a table, and enabling him to feed himself—all upon Jamie's spoken commands.

*National Aeronautics and Space Administration

Robots at School

If the robotics industry expands as rapidly as some analysts predict, many of today's elementary and junior high students will be needed as tomorrow's robotics experts.

Students at Willis Junior High School in Chandler, Arizona, are getting a jump on the future job market by participating in a pilot program—a separate "robotics class" scheduled right into their curriculum. If the class proves successful, other robotics classes may start popping up in junior high curriculums all across the country.

Some teachers introduce their students to computer graphics with the help of an 8-inch-high mobile robot, called the Tasman Turtle, that can draw with a special pen. Topo, B.O.B.'s look-alike "little brother," is another popular school robot. Topo is less expensive than B.O.B. because it lacks a built-in computer. Instead, Topo is radio-controlled from an external keyboard.

By using Topo, computer programming as well as robotics becomes easier to understand.

At one school in California, students learn programming by playing The Topo Olympics. The game's goal is to guide Topo safely through a complicated obstacle course. When Topo bumps into something, whoever is at the keyboard loses his turn.

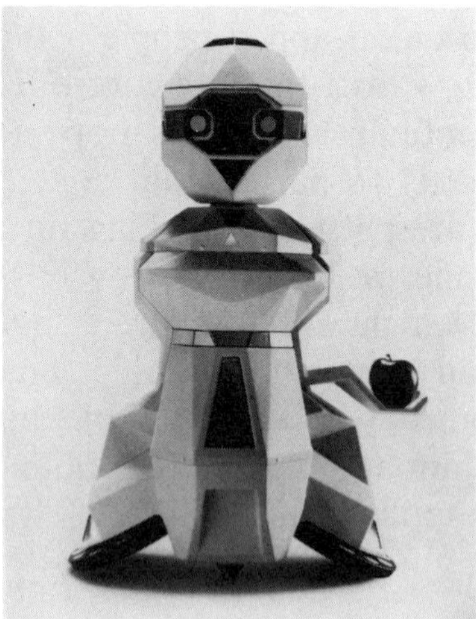

Photo courtesy of Androbot

Topo's a hit with students and teachers.

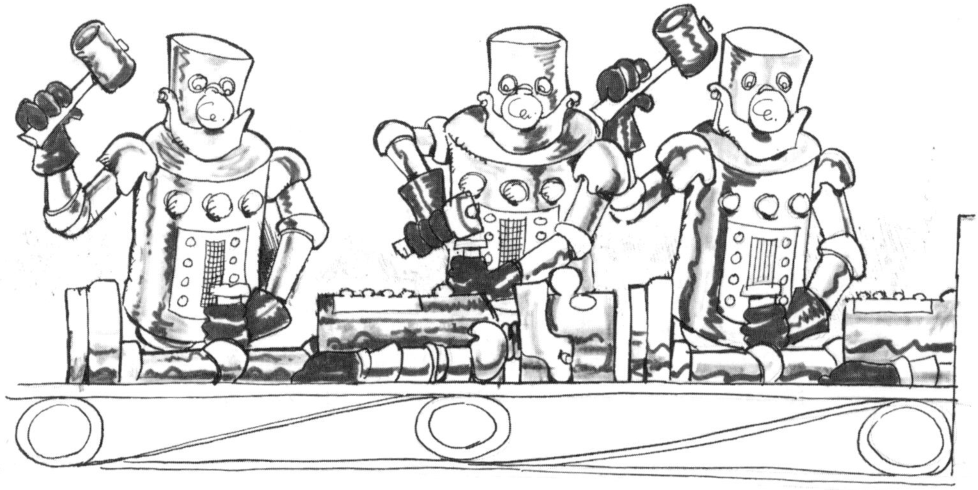

Robots Reproduce? / 8

The fictional robots in *R.U.R.* had two fatal flaws: They couldn't feel emotions and they couldn't reproduce.

Today's real robots may be programmed to *look* like they feel a certain way, but true human emotions are beyond them.

However, reproducing is not. At the Fujitsu Fanc. Ltd. factory near Japan's Mt. Fuji, under limited human supervision, industrial robots are already turning out replicas of themselves on the assembly line. We call these self-replicating robots.

Robot Families Planetwide

Dr. Georg von Tiesenhausen, assistant director at NASA/Marshall Space Flight Center in Alabama, envisions a time when self-replicating robots will be used all over Earth, as well as elsewhere in space.

"You could start with one factory of limited size," says Dr. von Tiesenhausen, "and by using self-replicating robots have not one, but twenty factories within a year."

Mobile robot systems—or families—could make it feasible for us to mine Earth's seabed for its rich deposits of manganese nodules, which are used in the manufacture of high-grade steel.

You would start this massive mission by lowering one robot into the depths. A year later you would have a few hundred machines combing the ocean floor. A few more years, and 500,000 robot underwater miners could be hard at work!

Meanwhile, troops of self-replicating robot well-drillers could spread out over Earth's deserts, converting today's wastelands into tomorrow's lush green crop-producing zones.

Assignment: Moon

You could send just one reproducing robot vehicle to the moon, and in a few years, swarms of robot rovers would be probing the entire lunar surface and radioing back to Earth what the moon's all about.

A human task force assigned the same mission couldn't complete it in less than a century.

And before humans settle on the moon in any significant number, self-replicating robots could build us domed cities with the weather and other living conditions engineered for optimum comfort and efficiency.

Energy for lunar settlements could be provided by robot families that build solar cells directly from raw materials on the moon. These solar cells could then be made into large solar arrays. The arrays not destined for moon service could be shipped to solar satellites, where they could be used to direct the sun's energy in concentrated beams to Earth.

Dr. von Tiesenhausen explains that a conventional solar cell factory might take 6000 years to make one million solar cells from the moon's raw materials. But a self-replicating system would get the job done in two decades!

Model of robot factory on moon that manufactures its own parts, as well as assembles them.

Cosmic Miners and Messengers

Other robot families could be called upon to mine the precious minerals known to exist in asteroids. These chunks of planetary materials revolving around the sun are found mainly between Jupiter and Mars.

And families of robot spacecraft built on the moon or on Earth-orbiting work stations may someday fan out after Pioneer 10 in an all-out effort to find other intelligence in our Milky Way Galaxy and beyond.

Model of a self-replicating lunar factory with parts already in existence.

Timetable for Self-replication

Dr. von Tiesenhausen sees two kinds of self-replicating robot systems possible. The first would begin with the parts it needs already available to it and would take about five years to become operational.

The second system would make its own parts from raw materials and then would start assembling itself. Scientists call this the "Santa Claus System," because it does everything they wish it would. This system would take approximately twenty years to put into production.

Robotics Wrap-up / 9

Robots are still in their infancy, and robot families that reproduce on their own are no more than twinkles in robot researchers' eyes. Yet the time is ripe for us to think about robots and the people who program and use them.

The future of robots—the good or evil they do—depends on what we, their teachers, instruct them to do. Hopefully, we will use them responsibly and wisely.

And then what? Once robots take over our most backbreaking and time-consuming labors, what will humans do?

Provided we have solved the unemployment problem in time, perhaps the human race will begin to concentrate its mental energies on new horizons. Maybe we'll even discover how to communicate with each other via ESP and save our vocal cords for our chats with robots!

People Are Precious

Once robots become better and better at imitating human beings, it is only natural for people to wonder just how special we are.

The answer to that question is that we're *very* special.

It is when we begin to consider the shortcomings of robots: their lack of courage or compassion, their inability to trust or respect, to emotionally grow, to laugh at a joke, or to see a rainbow and instantly have two dozen different happy thoughts go through their minds that we begin to appreciate how truly unique and wonderful human beings are—spirit and all!

The Deciding Factor

At the end of the *R.U.R.*, Alquist—the last human left alive on Earth—discovers Primus and Helen, two very special robots who can fall in love and reproduce.

"Go, Adam, go, Eve," Alquist says. "The world is yours."

In fiction, anything is possible. We can even imagine that someday we'll discover The Truth! The ability to build and then maturely use robots may be a major test to be passed before any planet gets invited to join The Galactic Society of Intelligent Life.

In reality, the robots are here. Their fate rests with us.

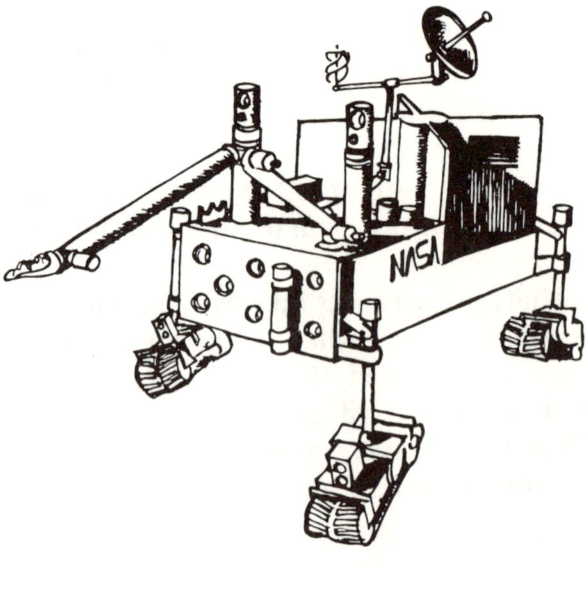

Glossary

Android: A robot in human form.

Artificial Intelligence: The ability of machines to think.

Artificial Skin/Tactile Sensor: Electronic device that allows a robot to know its environment by the sense of touch.

Asteroids: Small solid bodies ranging in size from a few miles to over 600 miles in diameter that revolve the sun in orbits, mostly between Mars and Jupiter.

Biochips: Single molecules of super fast, super small computer circuits that scientists theorize can someday replace the microchip. Biochips do not yet exist.

Byte: A string of eight electrical charges that can indicate a number, character, or symbol.

Chip: Another word for a tiny wafer of silicon with hundreds of thousands of microscopic electronic components etched upon its surface.

Circuit Board: A base for today's electronic circuits. It is usually made from a thin, flat piece of plastic or other insulating material.

Computer: An electronic device that processes and stores data.

Data: Information fed into and read out of a computer.

Electricity: One of the basic forms of energy on Earth. Electricity is carried by electrons and protons.

Electromagnetic Spectrum: Bands of different wave lengths from less than 10-trillionths of a meter long to more than 10,000 kilometers. Visible light occupies only a tiny band on the spectrum.

Electronics: A branch of science dealing with electrons and how they flow within such devices as electronic tubes and transistors. By controlling the flow of electrons, you can control the flow of electric current.

Exoskeleton: A metal frame humans can wear to give themselves extra muscle power.

Hot Cell: An insulated area in which radioactive materials are handled.

Humanoid: A robot having somewhat human characteristics.

Infrared Wavelength: Long light wavelengths beyond the visible color red on the electromagnetic spectrum. The naked eye cannot detect infrared wavelengths.

JPL: Jet Propulsion Lab.

Keyboard: A set of keys on a typewriter or computer.

Memory: The part of a computer where data is stored.

Microprocessor: A small integrated circuit that contains the complete central processing unit for a small personal computer.

Organic Molecule: Molecule containing one or more carbon atoms, whether of natural origin or produced in a laboratory.

Photoelectric Cell: A device that changes rays of light into currents of electricity.

Program: A list of instructions telling a computer what to do and how to do it.

Remote-controlled: Operated from a distance by a human.

Reskilling: Teaching a new skill to a worker whose previous skill has become obsolete.

Robotization: The replacement of human workers by robots.

Satellite: A natural or artificial body that revolves around a planet.

Sensors: Electronic devices on a robot that allow it to mimic human senses. A sensory robot is one with one or more sensors.

Showbot: Objects that look like robots, but are built to perform rather than work. Most are remote-controlled.

Sidelined: Moved from active participation to passive on-looking.

Slave/Master Arm: A device powered by electricity with a metal worker arm copying the movements of a control arm that is worn by a human operator.

Software: Programs, operating instructions, and other paperwork related to the operation of a computer.

Solid State: Refers to electronic components made from silicon or other solid substances like it.

Speech Synthesizer: A device that lets a robot convert electronic pulses into human speech.

Synthesize: Combine various parts into a whole.

Ultraviolet Wavelength: Short wavelengths in the electromagnetic spectrum that are beyond the range visible to the human eye.

User Friendly: Easy for a human to program and operate.

INDEX